War Bots

U.S. Military Robots Are Transforming War in Iraq, Afghanistan and the Future

By David Axe

Cover by Steve Olexa

NIMBLE BOOKS LLC

ISBN-13: 978-1-934840-37-5

ISBN-10: 1-934840-37-8

Copyright 2008 David Axe

Version 1.0; last saved 2008-08-13.

Nimble Books LLC

1521 Martha Avenue

Ann Arbor, MI 48103-5333

http://www.nimblebooks.com

Photographs by David Axe are © 2008 David Axe.

Cover design © 2008 Steve Olexa.

Other photographs are provided courtesy of Bryan William Jones, Department of Defense, Boeing and Northrop Grumman.

This book was produced using Microsoft Word 2007 and Adobe Acrobat 9.0. The cover was produced using The Gimp. The body text is Constantia, designed by John Hudson for Microsoft. The headings and captions are Consolas.

Contents

About the Author v
Introduction vi
Bomb-bots Go Boom 1
War Toys 5
Together at Last 8
TMI, Robot Dude 12
Ready, Set, Go, Crash! 15
Follow(er) 19
Man's Best Robo-Friend 23
"Trust Us" 25
Robot Flocks Darken the Skies 30
Spy Plane, Minus the Spy 33
The All-Seeing Robotic Dragonfly 37
There's No "I" in "Robo-Team" 39
How Long a Leash? 43
Navy's Secret Robot Spies 45
The Homecoming Queen is a Robot 49
Guerrilla Drone Warriors 51
Predators Evolve 53
Robots to the Rescue 55
"Kill Chain" 57
Killer Bots 60
Swarming 63
Build Your Own War Bot 65
These Legs Were Made for Thinking 67
Swimming Lessons 72
Postscript 74
Other Titles about Modern War from Nimble Books 75

Publish A "Nimble" Book 76
Letter from the Publisher 78

ABOUT THE AUTHOR

David Axe is a freelance journalist based in Washington, D.C. He has reported for *The Washington Times,* C-SPAN and BBC Radio from Iraq, Afghanistan, Lebanon, Chad and Somalia. He is the author of the *War Fix* series of graphic novels and *Army 101,* a nonfiction account of ROTC in wartime. He blogs at www.warisboring.com.

Introduction

What is a robot? Not everyone agrees.

If it's man-made, automatic and, in the words of one roboticist, "has some physical effect on the world," it's a robot, right?

But how much ability must it possess? How smart does it have to be—and how independent of human control? If you loosely define the term, then robots are everywhere. They're sliding doors, elevators, cruise controls, VCRs and microwave ovens.

But if you define them as self-contained, self-propelled and at least somewhat self-directing, then robots are a bit rarer. And by that definition, by far the biggest user of robots is the U.S. military. Only the military has the money, manpower, technical expertise and pressing need for large numbers of robots.

"Robotics can do three things for the future Army," Dr. M. Franklin Rose, an engineer working with the military, told The New York Times. "[Robots] keep soldiers out of harm's way, do the laborious and boring tasks and keep going long after a soldier is exhausted. And they have no fear ... "

Military robots come in all shapes, sizes and modes. They fly. They hover. They swim and dive. They walk, roll and crawl. Some carry weapons. Most do not. Nearly all of them come equipped with sophisticated sensors for peering at the world around them. Some are highly autonomous, restrained only by our deliberate decision to insert human-controlled switches into their behavior. Others, by design, are quite dumb. These are meant to perform simple tasks. Some are even meant to be disposable and cost just a few thousand dollars. One, Global Hawk, costs around $100 million per copy. All are intended to make warfare safer and easier for the user, and more dangerous for his enemy. They can do this because they have no fear.

Arguably the first military robots were guided bombs that made their debut in World War II. Modern precision-guided munitions, tens of thousands of them, still account for the majority of robots. But in the past decade large numbers of bots have entered service in other roles. Today they are our spies and sentries. They help find and defuse bombs. They loiter high in the sky or underwater, scanning for the enemy. Some carry weapons, but only fire them when ordered to do so by a human being.

In 2001, the Pentagon had just 200 robotic aircraft. In 2008 it had more than 5,000. The number of military ground robots jumped from 160 in 2004 to around 4,000 in 2006. Only underwater robots lagged: so far just a few dozen systems have entered service. Under the water is, after all, the toughest environment for robots. But even undersea bots will see a boost in coming years. The Pentagon has plans to spend at least $4 billion a year for the foreseeable future designing and building robots.

The spread of robots in our armies, navies and air forces has greatly advanced the science, engineering and techniques for mixing thinking people and thinking machines. And it has forced us to try answering a basic moral question. Just how much responsibility should we surrender to machines? If and when robots fulfill their promise to make war cheaper and easier for our side, will we discover that we wage war too lightly? Are we already guilty of that sin?

This book examines just a handful of the many types of war bots, and just a few of the ways they're being used in the expanding American-led "war on terror." Some of these robots have been in service for years. Some are still just prototypes. Between them they span the entire range of military robotics. Some are killers. Others are helpers. All of them are soldiers with no fear.

Figure 1. An Air Force bomb squad technician practices with his Talon ordnance-disposal robot at a base in north-central Iraq in 2005. Talon is one of the most popular bomb-bots. Army photo.

Figure 2. A Talon robot—note the arm and cameras—at a Navy training exercise in Djibouti in 2005. The Navy is in charge of developing bomb-disposal tactics for the entire U.S. military. Navy photo.

Bomb-bots Go Boom

The field of bomb disposal has benefited the most from military robotics. It's dangerous work finding, examining, disabling and destroying enemy bombs. More and more that job falls to a wide range of small robots.

Talon, built by Foster-Miller, is the standard bomb-bot for explosive ordnance disposal. The 100-pound, $100,000 Talon boasts cameras and an articulated arm and claw for probing suspected bombs—and for placing strips of C4 explosive to destroy them.

Talon and its handlers ride in special armored trucks. There's a ramp for loading and unloading the robot. The operators can remain inside under protection or guide Talon from outside. Either way, they rely on a radio console that includes video screens showing everything that Talon sees.

Unlike many flying robots, which can navigate on their own, Talon has very little autonomy. It's essentially a remote extension of its operator's eyes and hands.

There's even a machine-gun armed version of Talon called Swords. The U.S. Army tested several Swords in Iraq in 2007, but soon confined them to base over concerns that the Swords weren't reliable … and might shoot at the wrong targets.

Talon has proved extremely tough. Some Talons have survived multiple bomb blasts. Talons and other ground robots that are damaged in explosions can be hammered back together at "robot hospitals" situated on bases in Iraq and Afghanistan.

Figure 3. An Air Force Talon bot, nicknamed "Johnny Five" after the robot hero of the movie "Short Circuit," is nearly destroyed while clearing a roadside bomb near Kirkuk, Iraq, in 2006. Army photo.

Figure 4. Bomb-squad robot operators take cover as a roadside bomb destroys their Talon robot near Baghdad in 2006. The operators are flanked by specialized armored vehicles that all bomb squads ride in. Air Force photo.

Figure 5. A Navy mechanic works on bomb-disposal bots at a "robot hospital" near Baghdad during a sand storm in 2005. Robot hospitals in Iraq and Afghanistan "treat" hundreds of damaged robots every year. Army photo.

War Toys

Bomb squads are among the pioneers of modern robot tactics. After robots proved themselves in EOD work, they quickly spread to non-specialized combat units. MarcBot was one of the earliest small robots for everyday soldiers.

MarcBot is adapted from a commercial radio-controlled toy car and weighs just 25 pounds. Army units carry MarcBots in special tough suitcases. An operator steers MarcBot with a radio console. Like most ground robots, MarcBot has cameras and an arm for placing C4 explosives.

While useful for scouting around corners or investigating potential booby traps, the $10,000 MarcBot has big limitations. Because it's wheeled and not tracked, it's limited to relatively smooth surfaces. And early models quickly burned through batteries.

In the future, small tracked robots with better batteries will replace MarcBot. In 2007 Army squads began testing these upgraded PackBots from manufacturer iRobot, which also makes the Roomba vacuum-cleaner bot. The controllers for these test models were modified video game controllers, which Army engineers chose in order to take advantage of young soldier's gaming skills.

Some veteran bomb-disposal experts have pointed out the dangers in giving robots to more soldiers. There is a tendency among robot-equipped units to try clearing bombs on their own rather than call in the ordnance specialists. In other words, to have a fearless bot on their side can make soldiers cocky.

In January 2005, a 1st Infantry Division engineer unit in Baqubah, Iraq, for instance, tried to clear out insurgent bombs using a MarcBot. The robot's battery died. Rather than wait for an EOD team, the engineer captain walked right up to the bomb as though he were a fearless robot himself ... and just kicked it apart.

Figure ... Division ... prepare a MarcBot for investigating an Improvised Explosive Device near Baqubah, Iraq, in January 2005. Photo by David Axe.

Figure 7. The MarcBot approaches the bomb ... only to run out of power. An Army captain would then walk up to the bomb and kick it apart. Photo by David Axe.

TOGETHER AT LAST

Even "a twig in the road is an obstacle" for ground robots like Talon and MarcBot, according to one Army researcher. That means designers pay a lot of attention to robots' physical bodies—the means by which they roll, walk or crawl across the ground.

In the Army's main robotics center near Detroit, workshops are crowded with experimental robots sporting tracks, legs, wheels ... or combinations of all three. But the real focus here is on robots' brains: the algorithms that govern their behavior, the software that lets them think and the interfaces that allow human operators to get involved in both.

The problems and solutions vary, depending on the size and role of the particular robot. In 2006 Army engineer Terry Tierney was working on robotic cargo vehicles. His work applied to any robot that needed to travel long distances alongside other bots, or with manned vehicles.

Pentagon experiments such as the 2007 Urban Challenge robot race have showed that it's possible for robots using laser sensors to spot and tail each other. Tierney was trying to speed up and smooth out robot convoys to the point where they weren't just possible, but actually useful. Only then could the Army launch a program to actually build large numbers of robot cargo vehicles.

The benefit? They would take people off of dangerous roads in Iraq and Afghanistan.

One key is smarter blending of human and robotic decision-making. "What we have seen over the last five or six years in the [robot] community is significant advances in their ability to perform semi-autonomously," Dr. Thomas Killion, the top Army scientist, said in 2006.

But this isn't a perfect capability. "On occasion," Killion admitted, a robot "might have to call back and say, 'I'm stuck here and I need help or need to know if this is the right path.'"

What Killion was saying is that robot brains still aren't quite up to snuff. Most bots still need people's help. That's where Tierney came in.

"We're looking to address the 360-degree situational-awareness problem, to get robots' data into manned vehicles," Tierney said. Human operators could then see everything a robot was seeing and help the robot make decisions.

Army engineers demonstrated this using a pair of experimental control consoles. An engineer activated three large side-by-side touch-screens. One displayed a digital map dotted with icons representing friendly vehicles. Another display was fed by the command vehicle's own cameras while a third showed what a selected drone was seeing.

The "crewman" touched a robot's icon to bring up its external view on a screen. He could then control that robot and its guns and cameras using the same controller he was using for the vehicle he was in. Or he could tell the bot to operate autonomously. And when it needed his help, the bot would send an instant message ... and a window would pop up with the command options: go here, look there, shoot this. The human handler would just tap the command he wanted, and send the robot on its way.

The Army is planning to install this sophisticated robot controller in new armored vehicles as part of its $200-billion "Future Combat Systems," a family of robots and armored vehicles that's supposed to enter production in 2013.

Figure 8. Rough terrain is a major obstacle for ground robots, but the control software is just as complicated—if not more so. Here a bomb technician sends a PackBot to clear two rocket booby traps near a U.S. base in Afghanistan in 2007. Air Force photo.

Figure 9. Soldiers test a new PackBot in Texas in 2007. PackBot is manufactured by iRobot, the same company that makes Roomba robotic vacuum cleaners. Army photo.

Figure 10. An Army engineer demonstrates the Future Combat Systems robot control interface at a workshop in Detroit in 2007. Photo by David Axe.

TMI, Robot Dude

Even if you've built the perfect robot and the perfect control system, you still don't have a working robotics systems. The missing piece is the invisible radio link between the robot and its handler. The maximum available size of that link is called "bandwidth"—and it's a huge problem for the military.

With soldiers, civilian cell phone owners and commercial broadcasters all fighting for the same bandwidth, there's never enough to go around. That puts a lot of pressure on the military to keep down its own bandwidth appetite.

There's another benefit to this appetite suppression. Data overload between a robot and its human handler—"too much information," or "TMI," the kids might say—can overwhelm the handler. "You've got to balance sensor input and a soldier's needs," Bill Smuda, an Army engineer, said in 2006.

Smuda's projects included the speedy tracked PackBot, the four-legged Big Dog rucksack hauler, the combo tracked/legged Chaos search-and-rescue bot and Odis, a wheeled robot that uses a camera atop a collapsing "zipper mast" to peek into truck beds.

Most of these models have been deployed to war zones for trials. All are operated by just a single soldier using a handheld device—in some cases a modified Xbox gaming controller—including a video screen showing everything the bot is seeing.

"We must be aware of when [soldiers] get overloaded," said Gregory Hudas, one of Smuda's teammates. The Army intends a two-man crew to monitor as many as 10 robots, but Smuda recommended cutting that to just four.

At least one robotic system might actually alleviate other robots' burdens on soldiers, top Army scientist Dr. Tom Killion pointed out. "Some of the tools coming out of unmanned systems can be built into manned systems, in terms of autonomous driving capabilities."

He was referring to the robots in the Grand Challenge series of robot races, including Urban Challenge held in Victorville, California in October 2007. Technology from that race might soon find its way into a robotic autopilot for Army supply trucks.

Figure 11. Soldiers test out new PackBot robots at a training exercise in Texas in 2007. The Army is fielding tougher, longer-lasting bots controlled by modified video game controllers. Army photo.

Figure 12. Chaos, a search-and-rescue bot, can extend its tracks and use them as legs for climbing over rubble. The Army owns just a handful of the clever little robots for test purposes. Photo by David Axe.

Figure 13. An Army engineer shows off an Xbox controller modified to control an Odis inspection robot. The first soldiers to try out Odis asked the Army to make the robot "dumber." Photo by David Axe.

Ready, Set, Go, Crash!

The Urban Challenge race was barely two hours old when TerraMax ran into trouble. The 10-foot-tall autonomous robotic truck—the biggest by far of the 11 finalists in the competition—maneuvered into a parking lot on the 60-mile course at an abandoned Air Force base near Victorville, California.

The idea was to prove that it could park on its own. But something went wrong. The truck veered off the tarmac and nudged a vacant building.

As quick as that, TerraMax's race was over. The bot, built by Oshkosh Trucks, was ordered into impound, and over the next couple hours four more vehicles joined it, each disqualified for violating the race's fundamental rule: safety.

The obsession over safety wasn't just about dodging lawsuits. The ultimate goal of Urban Challenge—the third race in the Pentagon's "Challenge" series launched in 2004—was to advance technologies that might one day bump some soldiers off vulnerable supply convoys in Iraq and Afghanistan.

But that doesn't mean that future logistics patrols will be totally robotic. Mixed convoys will require robot vehicles to get along with human-driven vehicles on busy urban battlefields teeming with civilian vehicles and pedestrians. Preserving life wasn't a condition of Urban Challenge; it was the whole point.

TerraMax was the only Urban Challenge contender based on an in-service military truck. Other racers were based on civilian Sports Utility Vehicles. If any of the robots on display in Victorville that cold morning end up actually saving soldiers' lives in coming years, it most likely will be TerraMax.

While disappointing to the roughly dozen members of Team Oshkosh, TerraMax's disqualification actually had little bearing on the design's potential and development, said team member John Beck. "We want to perform well here, but ultimately we want to sell kits to customers."

As if Urban Challenge rules weren't stringent enough, Oshkosh heaped on its own requirements aimed at making the robot racer more "tactical." The result was a "near-prototype" vehicle that in quick heroic leaps blazed a trail for future fleets of robotic supply trucks.

Most of the Urban Challenge contenders shared a common design strategy. Teams combined a medium-size SUV with off-the-shelf optical and laser sensors tied to commercially available microprocessors running custom-designed software. All the racers relied on GPS for waypoints.

Teams differed in the number and mix of sensors they used, but most settled on spinning laser scanners with ranging radars or cameras as back-ups. Only TerraMax had cameras as its primary "eyes."

"There are no moving parts," explained Chris Yakes from Oshkosh. But there were drawbacks to optical sensors, too. Bad weather tended to blur the images.

While its sensor suites represented TerraMax's most important divergence, the vehicle's size was its most obvious departure. Some of the standard maneuvering algorithms used by many of the teams assumed a smaller robot. "Making all those algorithms work on a larger vehicle is a bit more complicated than on a smaller vehicle," Yakes said.

Anticipating these difficulties, Team Oshkosh built "dials" into its algorithms to allow it to alter some software parameters on the fly.

Considering this, plus the vehicle's robustness and the extensive testing preceding the race, Oshkosh was at a loss to explain TerraMax's minor collision with that building. Yakes speculated that there might be a software bug somewhere, but said the team would need to pore through a lot of data before it knew for sure.

Despite losing Urban Challenge—Carnegie Mellon University's modified Chevy Tahoe ultimately took the $2-million first prize—TerraMax in many ways was a more mature robot. Indeed, TerraMax's performance at Urban Challenge apparently validated the team's unique sensor and platform choices, and gave Oshkosh a solid rationale for continuing with development.

The next step, said team rep Katie Paulson, was to line up strictly military demonstrations that might attract the big government funding necessary to turn TerraMax into an actual weapon.

Up to this point, Oshkosh and the Army had pursued separate but parallel work on robot trucks. The Army's Convoy Active Safety Technologies, or Cast—using technology from a disqualified Urban Challenge race—earlier had demonstrated a compromise solution to the need for unmanned supply vehicles. The Cast experiment showed that robotic autopilots could take over from human drivers for much of a long-distance trip.

Oshkosh was hoping to follow up with trucks that drive themselves the whole way. "This story is not over," Paulson said.

Figure 14. TerraMax crosses the start line at the Urban Challenge robot race in California in October 2007. The Oshkosh-built robot supply truck was the biggest of the 11 contestants. Photo by David Axe.

Figure 15. TerraMax in impound following its minor accident during a parking exercise. Safety was the major requirement of the race, and even aggressive behavior could disqualify a contestant. Photo by David Axe.

Follow(er)

It's one thing to steer a ground robot in real-time using a hand-held radio console. It's much harder to design robots that can navigate themselves.

One method is called "following." A robot has sensors to keep track of a person or vehicle in front of it, and other sensors to watch the ground. Algorithms crunch the best direction for avoiding obstacles while also following the leader. In some follower-leader pairs, the leader also sends the follower Global Positioning System data via radio, as a backup to visual following.

Following came late to military robotics. The Army began dabbling in robot followers in 2001. It was six years before the Army was confident enough to launch a large-scale follower program. In the fall of 2007, engineers from sensor-maker PercepTek outfitted two logistics trucks with rudimentary leader-follower systems, named Cast, and tested them at Fort A.P. Hill in Virginia.

Both trucks had cameras on the front; servos under the dashboard for steering, accelerating and braking; and computers on racks in the back for crunching sensor data and operating the servos. The trucks took turns playing leader and follower. The driver in the front truck would flip a switch to leader mode. That told the truck to send GPS data to the truck behind it.

The driver in the rear truck flipped to follower mode. He pressed a big red button … then took his hands off the wheel. The truck now steered itself, attempting to stay behind the leader and on the road while also avoiding any pop-up obstacles.

How well did it work? PercepTek driver Peter Jarvis said that some of the braking was sloppy and the follower sometimes veered onto the shoulder. But soldiers who took turns "driving" the follower found that their ability to spot targets along the side of the road improved by 25 percent. With the robot driving, the soldier could focus on protecting the vehicle.

Besides improving supply convoy security, follower technology might result in robotic pack mules for combat units. The four-legged Big Dog robot mimics the shape of a pack animal. It carries up to 300 pounds of supplies on its metal frame. The first Big Dog test models were remote-controlled. Army engineers said the next step was to outfit Big Dog with sensors and follower algorithms so that the robot can accompany Special Forces on long hikes into dangerous territory, carrying their supplies so they can focus on staying alert for the enemy.

Future followers might include a wheeled "mule" robot that can carry hundreds of pounds of supplies for combat units. In addition to tailing soldiers into combat, these followers will also have the ability to autonomously navigate between forward positions and rear logistics areas.

Figure 16. Robotic armored vehicles test out "follower" technology in 2004. The same basic technology—including sensors and software—will help robots of all shapes and sizes follow each other and follow people. Army photo.

Figure 17. Two Marine Corps supply trucks fitted with PercepTek's Cast robot autopilots form a semi-autonomous convoy in Virginia in October 2007. Photo by David Axe.

Figure 18. A PercepTek driver takes his hands off the wheel, allowing Cast to drive the truck by itself, keeping in line with the truck ahead. Photo by David Axe.

Man's Best Robo-Friend

American infantrymen fighting in the mountains of Afghanistan have to carry everything they need to survive. Weapons, water, ammo, food, radios—a soldier might carry 75 pounds worth of gear, not counting his 30 pounds of personal body armor.

A hundred years ago the British Army in Afghanistan used mules to haul their gear, but the mules had to be fed, watered and negotiated with. What if you had a mule that was just as tough and nimble as the real thing, but could walk for a whole day without stopping and never flinched when bullets started flying?

That's the idea behind Boston Dynamics' Big Dog pack robot. "The goal ... is to create legged robots that mimic animal structure, mechanics and control to achieve animal-like strength, speed and mobility," the company said. In 2004, it unveiled a prototype with a stocky metal body, long thin legs and a stubby head with a camera and a laser scanner for eyes.

The Army quickly snapped up the Big Dogs for experiments, aiming to make them smarter with better sensors and software. In 2006 the Pentagon forked over $10 million to develop a Mark II version that can carry twice as much, twice as fast—up to six miles per hour.

But will the troops accept it when it enters combat around 2010? When Boston Dynamics unveiled the first Big Dog, more than a few engineers, soldiers and reporters called it "creepy." There was something about those mule-like legs and the bouncy, eager way Big Dog walked that was just weird and off-putting.

MIT robot designer Rodney Brooks called that effect "the uncanny valley," a term coined in 1970 by Japanese bot-maker Masahiro Mori. Brooks said human-like features—eyes, especially—are useful on robots for "communicating intent" to their handlers. But if a robot starts looking too human, or too animal, it triggers an instinctual revulsion. Designing robots with lifelike abilities is a delicate balancing act.

One design got it right. The Battlefield Extract Assist Robot, or Bear, built by Vecna, is meant for carrying injured soldiers out of combat. Its function is a direct duplication of a human action—carrying another person—so it needs human parts: hands, arms, legs.

To avoid the Big Dog's uncanny problem, Bear's designers gave it an exaggerated cartoony appearance. Sure, it has the general shape of a person, but its big round eyes and tiny ears make it look more like a giant teddy bear, according to the Army, which paid for development. Bear should be ready for field trials around 2012.

Figure 19. Robot designer Rodney Brooks and his teammates at MIT are working on bots with human-like facial expressions, such as "Domo," seen here. The idea is that facial "cues" might help people overcome an instinctual distrust of robots. MIT photo.

"Trust Us"

It flies nearly as high, stays up in the air longer, carries most of the same sensors and is cheaper to operate. But the Northrop Grumman Global Hawk spy drone is still years away from replacing its 50-year-old predecessor, the fabled U-2 spy plane. A major reason is that people just don't trust robots.

Despite multiple control redundancies and a proven track record over Iraq and Afghanistan, the Federal Aviation Administration won't let the 120-foot-wingspan Global Hawk fly in the same airspace as airplanes with actual people aboard.

The FAA's hang-ups over drones are purely psychological, according to Air Force Colonel Chris Jella, one of the first Global Hawk pilots, who control the big robots from trailers on the ground in California.

The technology and concepts for military robots are advancing faster than ever. But there's a flipside: people don't trust robots to do all the things that robots are capable of doing, for reasons that aren't always rational.

Six years after a General Atomics Predator drone launched a Hellfire missile to kill Al Qaeda mastermind Abu Ali in Yemen, human psychology has emerged as a major limitation in the continued advance of military robotics, to the extent that some amputees at Walter Reed Army hospital in Washington, D.C., have turned down the latest robotic prostheses because they don't feel comfortable with wearing a limb that "thinks."

It's a problem with robots throughout the military and civilians spheres, according to robot designer Rodney Brooks. "The [San Francisco] Bay area transit system can operate without a driver," Brooks pointed out, "but it ended up having a driver because people were more comfortable."

Case in point: In the early 2000s, some University of Utah students developed a fully autonomous, tile-shaped inspection robot called Odis. Surprisingly, the bot's autonomy turned out to be a problem for some Army operators who tested Odis overseas.

"Autonomous" means different things to different people, and there's no clear consensus on what exactly qualifies as "full autonomy," for at some point all robots at least communicate with human handlers.

The point is that Odis could go about most of its tasks without any human interaction. It could approach an idling vehicle, check under and around it and detect anything suspicious such as a bomb. But during initial tests, users complained, saying they wanted something they could control more, according to Army engineer Terry Tierney.

So Army engineers took the basic Odis and made it dumber, tethering it to a handheld remote control much like a kid's toy. Today the dumbed-down Odis is in military use in Iraq and Afghanistan and with law enforcement agencies in the U.S. But it's relatively manpower-intensive, more than it needs to be, all because of the discomfort of those early test operators.

These days, many engineers voluntarily limit the degree of autonomy they build into new robot designs, even if the technology enables a far higher degree of independence. EOD troops prefer relatively dumb robots that they control at every step from inside the protective shells of their armored vehicles. It's all about trust, and the lack thereof. For robots to make another leap forward, people have got to have faith in them to do everything they're capable of.

One way to build greater trust between soldiers and robots is to design bots that communicate with human-like body language. That's where Brooks came in. At his lab at MIT, he was working on design elements—some functional, some purely cosmetic—that might help build more trusting human-robot relationships.

"Suppose I want to show you how to perform some manufacturing task," Brooks said in 2007. "I come up to you, I put stuff in the center of your gaze and then maybe use little hand motions to indicate that this is what I want you to pay attention to. Then I glance at your eyeballs to make sure you're looking at what I want you to."

Brooks called these "intuitive social cues," and said robots should be able to mimic them. That would help put human handlers at ease.

It's all about gestures and gazes, Brooks said. Engineers could add structures simulating eyes and brows to a more autonomous bomb-disposal robot, and connect those to certain behaviors. In other words, before the bot reaches out its telescopic claw-arm to grab something, it makes a subtle "facial" gesture to announce its intent to reach.

The eyes don't need to appear very human. The gesture is more important than the eyes' actual appearance, Brooks said. Indeed, emotive structures that appear too human tend to have the opposite effect than intended: they cross into the "uncanny valley" of instinctual revulsion.

Emotive structures might help robots to win human trust. But there are limits to autonomy beyond those imposed by people's discomfort. "We'll never have [fully] autonomous vehicles because we've never had autonomous soldiers," Army engineer Bill Smuda said. "There is no soldier that just goes out by himself."

Figure 20. An Odis inspection bot at the Army robot workshops near Detroit in 2007. Soldiers demanded to have more control over Odis' actions. Photo by David Axe.

Figure 21. Despite its excellent track record, many government officials don't trust the Global Hawk spy drone to fly in the same airspace as manned airplanes. Northrop Grumman photo.

ROBOT FLOCKS DARKEN THE SKIES

In 2004 over Baghdad, a small spy drone collided with an Army Special Forces helicopter on its way to a top-secret mission. Luckily the helicopter was undamaged and no one was hurt.

The culprit? A five-pound Raven, hundreds of which buzz around Iraq at all hours. Even more Ravens are on the way.

Raven got its start in 1999 when the Army needed its troops on the ground to see over hills or around tall buildings. The solution? Take a four-foot-long Styrofoam model airplane built by AeroVironment and outfit it with cameras and a tiny computerized brain.

A soldier throws the Raven into the air, then taps Global Positioning System coordinates into his radio console, beaming them to the drone. When the Raven arrives at its destination, miles away, it transmits video back.

It was breathtakingly simple, and it worked. After an initial batch, the Army ordered more than 3,000 Ravens for troops in Iraq. The Marines ordered 500 of their own. Now Raven is the most numerous—and one of the most potentially useful—airplanes in the war in Iraq.

With all those Ravens buzzing about, collisions like the one in 2004 were bound to become a problem. And the Air Force didn't want to lose one of its $50-million F-16 fighters in a crash with a $30,000 robot. So in 2005 the Air Force asked the Pentagon if it could take command of all aerial drones, so that it could better coordinate their flights.

The Army, Navy and Marines cried foul, calling the proposal a power-grab. But the Air Force had a point. Ravens were really getting under foot, even causing problems with the radio jammers the Army uses to block roadside bombs from exploding. In 2006 soldiers complained that the jammers were confusing the Ravens' control signals, causing them to nose-dive into the ground. Fortunately, Raven's wings were designed to pop off rather than break, so the tiny drone usually survives the crash.

The Pentagon never allowed the Air Force control of the Raven, maintaining that altitudes below 3,000 feet still belonged to the Army and

Marines. This left the Army and Marines to tend to their growing flocks of Ravens, but it didn't solve the underlying problem. There's only so much sky, and more and more robots, airplanes and helicopters trying to fill it.

And while the Air Force's robot-directing power-grab might have been thwarted for now, the crowded skies have made many Army commanders reluctant to use their Ravens. With all the potential bureaucratic headaches resulting from robot usage, some officers now prefer to just send a man to scout for the enemy in place of a Raven. It's one of the unexpected backlashes against robots' rapid spread.

Figure 22. Army paratroopers inspect a Raven drone before hurling it into the air during a demonstration in North Carolina in 2008. Raven is the most numerous flying robot. Army photo.

Figure 23. An Army Raven operator assembles his drone before launching it on a search for roadside bombs in Iraq in 2006. The five-pound Raven has become so numerous that many commanders refuse to fly them over airspace-crowding concerns. Navy photo.

Spy Plane, Minus the Spy

On May 1, 1960, U.S. Air Force pilot Gary Powers was shot down over the Soviet Union in his U-2 spy plane. For two years he was held prisoner while, back home, the U.S. government worried that he was spilling the secrets of his high-tech jet.

After the Cold War, the demand for spy flights only grew, but nobody was willing to risk a Gary Powers incident over, say, Afghanistan. Luckily there was an easy fix. Keep the high-flying spy plane with its sophisticated cameras and radars, but lose the pilot. The result was Global Hawk, a $50-million flying robot with the wingspan of a 737 airliner, hand-built by Northrop Grumman.

Global Hawk usually takes off from some secret overseas base and spends 24 hours orbiting 60,000 feet over enemy territory, beaming images back to operators sitting in a trailer in California. If Global Hawk gets shot down, the operators hit a self-destruct button to keep its tech out of enemy hands. The self-destruct works so well that during one 1999 test, technicians accidentally blew up the second prototype in mid air.

Global Hawk was still just a prototype when the U.S. invaded Afghanistan in 2001, but the Pentagon couldn't wait to send it into battle. There were still bugs to work out, and half of the prototypes sent to war over the next five years crashed and burned. Still, Global Hawk was a hit—just two of the huge birds could keep round-the-clock surveillance of all of Iraq—and since those early flaws were fixed not one has crashed.

But that hasn't convinced the FAA. Global Hawk flights over war zones are unrestricted, but the FAA still limits domestic Global Hawk sorties to certain corridors around major airbases. If the Air Force wants to fly the drones anywhere else over the U.S., it has to apply for permission. Military officials have objected, pointing out that every Global Hawk has several pilots monitoring it at all times. Sure, they might be thousands of miles away, but they're still pilots. And they're accountable if anything goes wrong with their robot.

Unlike with many other aerial drones, which are steered by a pilot with a joystick and a forward-looking video feed, Global Hawk navigates itself to pre-set coordinates. Global Hawk pilots are managers, auditors and insurance—not just stick jockeys. Pilots' roles are similar aboard many modern airliners, which are packed with robotic systems.

So will Global Hawk's unprecedented autonomy pave the way for robotic airliners? Technologically speaking, maybe. But the biggest problems aren't technological—they're psychological. Because people don't trust them, for the foreseeable future robots won't stray far from human supervision. And for that reason, the military is looking at ways to better combine people and robots into seamless teams.

Figure 24. In 2007, a brand-new Global Hawk departs the Northrop Grumman factory in California. The Air Force plans to buy around 50 of the giant robots. Counting sensors, each one costs nearly $100 million. Northrop Grumman photo.

Figure 25. Air Force mechanics elevate a Global Hawk in order to test its fuel pumps at a facility in California. Half of the Global Hawk prototypes crashed while flying missions in Afghanistan and Iraq, but later models are much more reliable. Air Force photo.

Figure 26. Pilots fly Global Hawk from trailers in California. Control signals are beamed to the robot thousands of miles away by satellite, using one of the dishes seen here in 2006. Photo by David Axe.

Figure 27. A Global Hawk snapped these photos of California wildfires on behalf of firefighters in 2007. The FAA limits when and where drones can fly in U.S. airspace. Air Force photo.

THE ALL-SEEING ROBOTIC DRAGONFLY

Like a giant bug, it hovers behind cover, patiently waiting for its enemies and zipping out to attack. Fire Scout, built by Northrop Grumman, takes a lightweight Schweizer 333 civilian helicopter, slices out the cockpit and adds sensors, computers, radios and weapons stations.

The result is a $15-million military-grade drone chopper that can take off from any clearing, hide behind trees and hilltops, hover in front of doors and windows and fire guns and laser-guided rockets up the tailpipes of fleeing cars. The Army is planning to buy hundreds of Fire Scouts. The Navy wants its own versions to help fight pirate boats and suicide bombers in shallow coastal waters.

Fire Scout might join law enforcement, too. Due to their maneuverability compared to airplanes, helicopters are already a favorite of police departments everywhere. Now imagine your local cops equipped with a robot chopper that can stay in the air longer, doesn't need a lot of human supervision and can be fitted with weapons.

The good news is that just because a drone can be armed, doesn't mean it will be. The military is mostly interested in Fire Scout for its ability to take off in tight spaces and track bad guys, not its ability to kill them. There are even rumors that U.S. Army Special Forces bought a robo-chopper similar to Fire Scout, but weaponless, for carrying (living) prisoners of war into detention.

Figure 28. A Fire Scout test fires rockets during a 2005 test in Arizona. The Army and Navy are interested in outfitting the robot chopper with weapons, but its major mission is surveillance. Northrop Grumman photo.

There's No "I" in "Robo-Team"

Modern battlefields are dangerous places for helicopter crews. Sometimes the first sign of danger is a rocket-propelled grenade streaking into the sky. For that reason the Army wants to team up choppers and robots. "We're trying to use robots to be point men," Army Colonel Charles Bush said in 2007.

Drones trailblazing for helicopters might help the military avoid the kinds of fatal shoot-downs that U.S. forces in Iraq suffered in early 2007. The Army and Marines lost nine helicopters in a six-week period in January and February.

The main aircraft for these robot-chopper teams are the Fire Scout and the latest version of Boeing's Apache attack chopper, although other drones and choppers are also candidates. The basic hardware for tethering Fire Scouts to Apaches is ready now, but the unique software still needs work, and so do the teaming tactics. Army officials said that these might require years of experimentation.

The Navy, which is also buying Fire Scouts, has similar plans to team robots and manned aircraft, including helicopters. Again, experiments have proved the basic hardware and concepts, but widespread use hinges on improving the software. And software tweaks depend on procedures that, like in the Army's case, might need years.

The Army first proved the concept of robot-chopper teaming beginning six years ago, according to Larry Plaster from Boeing. The company had an Apache crew control all aspects of a drone's flight except takeoff and landing, which were handled by a ground crew. "We finished that up in 2004," Plaster said. "Then we gave the aircraft to the Army. They flew it around for a year, taking it to various units to show how it worked." But after that the modified Apache and its drone buddy were parked while industry kept working on the control software and the Army pondered the tactics.

In the meantime, Boeing took lessons from the test and applied them to the latest Apache, due in 2010. According to Plaster, the hardware is pretty straightforward, just as the basic hardware for ground robots is decades-old and thoroughly wrung-out.

But software for airborne control of drones has proved trickier than software for ground control of drones, due to the difference in tactics on the ground versus those in the air. "Conceptually, it's not really different," said Steve Petty, an Army civilian. "A pilot is a soldier. He's got a screen. He's got a certain number of sensors or platforms he's controlling. The difference is, he's flying." The big problem is with "integrating those controls with his flight software."

"You have to have a fairly robust software upgrade for UAV controls" in helicopters, Plaster says. In other words, without changes to the way people fly helicopters, it's just too much work to fly both your own chopper and a robot at the same time.

Figure 29. A Fire Scout angles in for the type's first landing at sea, aboard the amphibious ship Nashville in 2006. Fire Scout can stay aloft four around four hours. Navy photo.

Figure 30. Fire Scout is an example of a successful robot based on a manned vehicle. The robot chopper is a commercial light helicopter with the cockpit replaced by computers, radios and servos. Navy photo.

How Long a Leash?

The Army has to decide how much autonomy it will grant the drones in its future pilot-drone teams. For robots, autonomy is largely a function of software: algorithms govern a drone's behavior, instructing it when to ask human operators for commands.

"The real problem with autonomy is … I think military guys want to have control over everything they have in their battlespace," Army researcher Steve Petty said. "They don't trust autonomy."

"Nobody runs around with total autonomy," said Colonel Steve Bush from the Army. "If you do, you run into things or get shot. You've got to know where all your friendly assets are. It's an airspace issue. You're not going to have UAVs running around looking for targets [fully autonomously]. They may operate somewhat autonomously within a certain area."

Defining that "area"—what a chopper-controlled drone will be allowed to do on its own—takes plenty of experimentation, Bush said.

The Navy is in a similar boat, according to Mike Suqua, from Northrop Grumman. "The Navy is interested in manned-unmanned teaming with its H-60 [helicopter] fleet. But we probably won't get to that until we're through some hoops."

Once all the wrinkles are ironed out, unmanned Fire Scouts and manned choppers might form flexible hunter-killer teams, with both platforms equipped to spot and kill targets. But arming Fire Scouts remains just a possibility for both services. "Both have 'desirement,'" Suqua said. "I've heard various people in the Army at various levels of responsibility say they want the option to arm [drones]. We went out and demonstrated the capability to shoot rockets on Fire Scout as a proof of concept. I know the Navy has been very interested."

Figure 31. Small robots such as this prototype hovering drone that's part of the Army's Future Combat Systems tech incubator, will team up with manned vehicles to form scout and hunter-killer teams. Army photo.

Navy's Secret Robot Spies

While the Army and Air Force have added thousands of robots in recent years, the Navy and Marines have been criticized for being slow to embrace unmanned vehicles. There are two major reasons: one is that nobody has built a high-performing robot yet that can handle the stress of flying off an aircraft carrier. The other is that the Navy's and Marines' major milieu—the water—is a very difficult environment for robots. Especially under the water.

But the sea services do have a small number of aerial drones, including Scan Eagles built by Boeing, and have apparently used them to snoop for suspected extremists in Somalia. In early 2008 Somalis reported finding a crashed drone in the remote southern part of the country, at around the same time that the Navy had launched a cruise missile to kill a suspected Al Qaeda sympathizer.

"It's a small unmanned American plane," said Mohamed Mohamoud Helmi, a government official. "It was flying from the direction of the ocean and it crashed in an area where children were playing football. It doesn't appear to be damaged and we are ready to hand it back to anyone who claims ownership."

"I saw the small plane, it's about one meter and a half," one resident said. "It has cameras on it and things like computer components."

Based on the description, the drone appears to have been an Insitu Scan Eagle—a 40-pound bot launched by catapult and recovered in a net. The Marines use leased Scan Eagles for spotting insurgents in Iraq. The Navy recently announced it would buy some of the robots outright, and has tested Scan Eagles aboard cargo ships, destroyers and Special Operations boats.

The latter is key, considering the Navy's cruise missile strike in Somalia. While the crashed Scan Eagle might have been used to conduct post-strike reconnaissance, it's more likely the drone was supporting U.S. Special Forces on the ground.

Good, high-fidelity intelligence is hard to come by in Somalia. The lack of info practically begs for Special Forces teams on the ground to provide targeting data. There are persistent rumors that American special operators were in Somalia in 2006. Perhaps they remain. The Scan Eagles, launched from boats or warships off the coast, might be the commandos' eyes in the sky.

Figure 32. The Marines use rented Scan Eagles, flown by Boeing and Insitu employees, to scout for insurgents in western Iraq. The Navy is planning on buying some for shipboard use. Boeing photo.

Figure 33. An Australian Scan Eagle is launched by catapult for a mission over Iraq in 2007. For years Marine commanders have begged for more Scan Eagles to keep watch over western Iraq, but the Marines have been slow to rent or purchase more. Air Force photo.

Figure 34. Contractors in a Special Operations boat recover a Scan Eagle following tests off the coast of California in 2008. Scan Eagles are launched by catapult and recovered in a net. A drone crash in Somalia perhaps indicates U.S. commando activity in the country. Navy photo.

Figure 35. A contractor controls a Scan Eagle from a workstation in Iraq in 2007. Instead of rapidly building up their own drone forces, the Navy and Marines have relied on civilian contractors and rented robots. Air Force photo.

The Homecoming Queen is a Robot

The hottest airplanes in the United States Air Force in 2008 weren't factory-fresh $150-million F-22 Raptor fighters with their stealthy airframes and super-cruise ability. They weren't the $1.2-billion B-2 Spirit bombers that were getting new radars and data-links. They weren't the new F-35 Lightning II fighters that had just begun flight testing and might cost as much as $90 million per copy.

No, the most-requested airplanes in the USAF were bulked-up remote-controlled designs costing less than $20 million. The 2.5-ton MQ-9 Reaper, cousin of the half-ton MQ-1 Predator, attacked its first target in southern Afghanistan in October 2007, around six years after the Predator fired its first missile in combat.

The snub-nosed robots with the pusher engines, manufactured by General Atomics, have become favorites of soldiers, Marines and commandos in Iraq and Afghanistan. They can loiter for hours with loads of missiles and bombs, scanning the terrain below, waiting to attack the bad guys.

Predators and Reapers aren't just popular with the ground troops. They represent a smart investment for an air force that is struggling with inadequate budgets and a large, aging fleet left over from the Cold War.

Back then, fighter duels over central Europe were a main mission. Today it's tracking and killing elusive insurgents in teeming cities and, increasingly, in remote and sparsely inhabited regions of the world. For countering insurgencies, an Unmanned Aerial Vehicle is a better weapon than any large, supersonic fighter jet. A UAV's advantages include the ability to remain aloft for much longer, the ability to change crews mid-mission, and the almost unique ability to remain stealthy while performing the mission.

But Predator and Reaper aren't without their naysayers, and there are big challenges in taking advantage of their unique potential. Despite their excellent records in recent years, the types suffer from inadequate cockpit design and lingering problems mixing robots and manned airplanes in the same airspace.

Figure 36. A Reaper taxis out for a mission over Afghanistan in 2007. British troops called the armed robot a "mini A-10," referring to the legendary manned attack aircraft with the tank-killing 30-millimeter gun. Air Force photo.

Figure 37. A Predator lands at Creech Air Force Base in Nevada in 2008. Predator has been armed since 2001, and has been used to assassinate several high-profile terror suspects. Photo by Bryan William Jones.

GUERRILLA DRONE WARRIORS

At a desert airbase outside Las Vegas, Nevada, a guerilla-style team of officers and enlisted airmen is trying to solve the biggest problems with current flying drone designs, while also flying nonstop combat missions using aircraft that are located thousands of miles away.

Creech Air Force Base, once a mostly abandoned relic of the Cold War, is now the central hub for Predator and Reaper operations, making it arguably the most important base for the future U.S. Air Force.

The 432nd Wing trains all Air Force Predator and Reaper crews, as well as crews from the U.S. Navy, the Customs and Border Service and the British Royal Air Force. The wing's operational squadrons deploy aircraft, maintainers and launch crews to the Middle East and Central Asia, among other destinations, while the pilots and sensor operators remain at Creech, controlling the Predators and Reapers during missions that last longer than half a day.

The Air Force is looking to expand its combat-tasked Predator and Reaper force from roughly 15 round-the-clock "orbits" in 2008 to at least 21 by the end of 2009. Each orbit comprises three or four robots. Around 50 Predators have been destroyed in crashes, most of them during the type's early years, leaving some 100 Predators and 10 Reapers in U.S. service. Another 150 of the medium drones, built by General Atomics, are planned through 2010.

Predator and Reaper missions are all about teamwork. The aircraft themselves are usually housed overseas alongside a ground control station and around 50 people. There are another 30 people and an additional station back at Creech. Airmen in the deployed station launch and land the plane, passing control via satellite signal to the Creech-based crew for the combat phase of the mission.

The control trailers at Creech are unimpressive on the outside; on the inside they look like something out of science fiction, with banks of monitors relaying everything the Predator or Reaper sees with its cameras and, for the most advanced models, radar that can "see" through clouds. Orbits are positioned by regional commanders in places where they can most efficiently peel off to aid combat troops.

In 2007, Creech crews sent drones to assist ground troops on average once a day and flew in support of pre-planned raids twice a day. They fired Hellfire missiles once every three days and, beginning in October when Reaper reached the field, dropped 16 laser-guided bombs. That year USAF Predators and Reapers flew 83,000 flight hours. In 2008, the Air Force expected to fly more than 90,000 Predator and Reaper hours and said it would add satellite-guided bombs to the Reaper's arsenal, according to Colonel Chris Chambliss, the 432nd Wing commander.

Figure 38. A Reaper over Creech Air Force Base in 2008. The 432nd Wing at Creech is equipped with dozens of armed drones and deploys robots and launch teams all over the world, but the pilots stay behind, operating the bots by satellite link. Photo by Bryan William Jones.

PREDATORS EVOLVE

Air Force drone operations are changing fast. Early-design Predators made the type's combat debut flying unarmed reconnaissance missions over the Balkans in 1996. New models are more reliable, more powerful, carry improved sensors and, most notably, are wired for under-wing ordnance.

But one aspect badly in need of greater improvement is the control station. Lieutenant Colonels Mark Hoehn and Greg Christ, from the robot-equipped 432nd Wing, say that the banks of monitors don't give pilots a view wide enough to ensure confident control in crowded airspace. They call the current view a "soda straw."

Creech crews also want more adaptive controls. In a manned fly-by-wire fighter such as the F-16, the control stick is programmed to be less sensitive during landings, allowing for more "slop" in the control stick when air is moving over control surfaces more slowly. Predator and Reaper lack this flexibility, and tend to get twitchy on takeoff and landing. Crashes near airfields have accounted for many of the roughly 50 Predator write-offs. The crash rate has declined greatly in recent years as operators gain experience.

With "cockpit" improvements, the Air Force might be able to reduce the crash rate even more. Not only will this save money, it will reduce the likelihood that crashed drones might give away sensitive missions. For every publicly-acknowledged mission coming to the rescue of pinned-down Marines, the drones also fly hush-hush sorties in politically-sensitive environments.

In 2001, a CIA-operated Predator flying over Yemen launched a Hellfire that killed Abu Ali, the mastermind behind the USS Cole bombing that killed 17 American sailors. And in February this year, a Predator or Reaper reportedly entered Pakistani airspace in order to fire a missile that killed 12 suspected Pakistani insurgents.

Such a mission would be unthinkable for a manned fighter. Only a drone like Predator is capable of such stealthy infiltrations of countries where the U.S. won't dare execute obvious military attacks. Predator and Reaper are quiet and small enough not to be widely noticed by volatile local populations, and have the long loiter time necessary to carefully verify targets before firing—both requirements of governments, like Pakistan's, that are officially allied with the U.S. but fear popular backlashes.

The flexibility to fly surveillance and attack missions and pull off secretive missions, combined with low acquisition costs, makes the Predator and Reaper ideal for the 21st-century Air Force.

Figure 39. A Reaper spins up for a mission over Afghanistan in 2007. This evolved version of the Predator can carry almost as much ordnance as an F-16 fighter—and can loiter for far longer. Air Force photo.

ROBOTS TO THE RESCUE

Predator and Reaper drones just might be the salvation of a military service that faces a perfect storm of rising costs, shrinking budgets and changing strategy.

The U.S. Air Force is in trouble. The rising cost of high-tech jets, the people to fly and maintain them and the gas to make them go threatens to put the service "out of business," according to former Air Force Secretary Michael Wynne. The average age of Air Force planes is now 24 years, versus just eight years in 1967.

Are robots a solution to the Air Force's problems? They're cheaper than, but just as effective as manned aircraft, according to proponents. Critics say robots with remote operators aren't anywhere close to achieving the creativity and flexibility of human pilots in cockpits, and that moreover drones lack the speed and payload capability to keep up with a wide range of missions and threats. Wynne, back when he was a Pentagon acquisitions official, said manned fighters would dominate for decades. "Right now, our strategy is to go for an all-stealth, manned, tactical air fleet good through 2040," he said.

Some say drones will be ready sooner—or are ready now—to replace manned fighters. "Air power could be transformed by the increasing substitution of unmanned for manned systems," one think tank posited. "As it passes through this transformation, the Air Force will have to transcend the essence of its founding identity: manned flight. This will pose challenges not only to its core institutional culture, but also to its warrior ethos."

Predator and Reaper crews at Creech have risen to that challenge. They said their daily interaction with ground forces that is incumbent in robot operations makes them a tighter part of the front-line team, despite them being thousands of miles away. Creech's drone operators might sit in air-conditioned trailers, but they are still very much warriors.

They highlighted the advantages of drones, pointing out that an F-16 will burn through 300 pounds of fuel just starting up and taxiing, while a Predator will fly an entire 16-hour mission on the same amount of fuel. Reaper, for its part, soon will be cleared to carry 5,000 pounds of ordnance, roughly the same as an F-16's combat payload. Yes, current drones are slower than an F-16, but an emphasis on dash speed means the F-16 only has 10 minutes on station, however fast it gets there. A Reaper might hit a target just as quickly, if not quicker, because it has been orbiting nearby for hours, silently and patiently waiting.

Figure 40. An early Predator model receives maintenance attention after a mission over Afghanistan in 2002. As many as half of the early-build Predators were lost in accidents, but reliability has improved in recent years. Marine Corps photo.

"Kill Chain"

Of all the military services, the Air Force has the biggest fleet of armed drones. But it's the U.S. Army that's developing the technology and procedures to make drones, manned aircraft and people work together in seamless teams. In the future, Fire Scout helicopter robots and manned Apache attack choppers will fly side by side. But it was in Iraq in 2007 that the Army first paired up flying bots and other weapons in meaningful ways.

It was a move motivated by one compelling need: to reduce the time it takes to spot an enemy from the air, identity him and kill him. To attack elusive insurgents planting roadside bombs—the biggest killer of U.S. troops—the Army had to be quick.

They call this the "sensor-to-shooter chain" or "kill chain." Before, that chain might be hours or days long. Only in recent years has the Army come into possession of the airborne drones, data-links, helicopter cockpit displays, software and other systems that, assembled the right way, can take a few links out of the chain, reducing attack times. In 2007 the Army revealed a secret task force in Iraq, called Odin, that was killing insurgents within minutes of spotting them.

Army Brigadier General Stephen Mundt claimed Odin had killed more than 2,000 insurgents in 2007 alone. "Odin's after them, and taking them down in big numbers," he said.

In that year the major links in this short kill chain were: Shadow and Warrior drones, Apache helicopters, modified transport planes for relaying radio signals, new rugged ground terminals for soldiers on the ground and software for managing incoming data.

In May 2007, the Army showed off a video collage from Iraq that it said represented a recent short-chain attack. In the video, a suspected insurgent bomb-planting squad was spotted by a patrolling Apache that watched the bad guys using its infra-red sensors, trying and failing to identify them until it ran out of gas and had to return to base. The Apache crew beamed the video and target coordinates to soldiers in an operations center who promptly passed it all along to operators of a Shadow drone and a larger Warrior drone, a model based on the Air Force's Predator.

The two drones orbited the suspects and together verified that the men were placing an Improvised Explosive Device. And when a fresh Apache showed up, the drones handed back the target data via the operations center and the Apache fired a Hellfire missile. The drones continued orbiting until they confirmed that the men were dead.

The entire engagement lasted only minutes, but it's possible to do even better by adding better systems and including the Air Force, the Army said.

The inclusion of Air Force planes in the Army network belied serious disagreements about how the two services should cooperate on drones. The Air Force had been lobbying the Pentagon and Congress since 2005 to become the "executive agent" for all large and medium U.S. military aerial drones. One goal was to ease airspace crowding by having one agency in charge.

But Mundt said the two services worked at different levels, so giving all robots to the Air Force was a bad idea. "The Air Force is looking at the operational and strategic. The Army is all tactical. It's a matter of scale," he said.

Figure 41. A Navy mechanic works on a Shadow drone in Iraq in 2008. The Army and Navy work together on Shadow operations. The medium drone is a key part of the Army's "kill chain." Army photo.

Figure 42. A Shadow launches in Iraq in 2006. Shadows can pass target coordinates to other drones and helicopters in order to quickly identify and kill insurgents. Army photo.

Killer Bots

On November 3, 2002, a Predator drone flown by the CIA fired a missile at a car in Yemen, killing Abu Ali, the mastermind behind Al Qaeda's attack on the U.S. Navy destroyer Cole that killed 17 sailors in 2000. It was first reported assassination by a robot—and an early experiment in turning military drones, normally used for spy missions, into killers.

Of the thousands of robots in U.S. military service, just three types numbering only a few dozen, routinely carry weapons. These are Predator and its larger cousin Reaper, and the Army's Swords ground robot. For more than five years Predators have orbited over Iraq and Afghanistan, ready to pounce on terrorists and insurgents. The military doesn't track body counts, but Predator has racked up probably dozens of confirmed kills.

Swords, on the other hand, several months after its summer debut in Iraq hadn't even shot at anybody yet. Firing missiles from a remote-controlled airplane is one thing. Cutting loose a machine-gun-armed robot on the streets of Baghdad is quite another. The first Swords were built in 2004, but they never deployed because they had a tendency to spin out of control and react to their operators' orders ten seconds late—both habits that would've made them more dangerous to their human comrades than to the bad guys. But the 2007 version is much more reliable. One baby step at a time, the Army is giving its killer bots more freedom.

Not everyone is so happy with killer robots. A philosophy called "Transhumanism" argues that autonomous technology might one day destroy the human race. Some have called this the "Terminator argument," after the sci-fi film series where human survivors battle evil killer robots from the future.

Predator and Sword are pretty much the furthest thing from James Cameron's walking death machines. Both are controlled in real-time by an operator using a radio or satellite up-link. It takes three steps to fire Swords' gun ... and a long series of permissions from human bosses to launch Predator's missiles. Ironically, there are actually more people in the decision cycle for many of today's robots than for traditional tanks, fighter jets and artillery.

Still, the idea that you can sit in an air-conditioned office in Las Vegas, where all Predators are controlled, and fire missiles at people on the other side of the world, has got some people worried. When "cubicle warriors"—that's analyst Peter Singer's term—replace soldiers on the ground, warfare becomes a video game, but only for our side. The pain and suffering of our enemies, and of anyone caught in the crossfire, is still very real.

That's not really a problem right now. These days killer drones like Predator and Swords support and accompany human ground troops. But more autonomous killer drones are right around the corner. iRobot, makers of the Roomba robotic vacuum-cleaner, has designed a bot called Warrior X700 that is much more independent than Swords and can even think its own way around obstacles or right itself if it gets flipped over.

And this year the Navy launched a billion-dollar program to begin putting free-thinking robot fighter planes on aircraft carriers. In 2008 the X-47 Pegasus was just a prototype, but in 10 years it would be capable of taking off, spotting targets, dropping bombs and landing on a pitching carrier deck, all with minimal guidance. Pegasus and X700 are paving the way for future robotic warriors that could in theory put human soldiers and pilots entirely out of business—and in any event will make killing quicker, easier and cheaper for armies that can afford them.

But even the military has doubts about fully autonomous killer bots, and seems poised to artificially limit robots' capabilities. The Navy, for instance, said it would build human permission into Pegasus' firing procedure, even though it didn't have to. And the Army is putting the brakes on robot autonomy and is actually planning on increasing the number of human infantry in its combat brigades, despite the huge growth in the robot force. The military seems to realize that just because a robot can do a thing, doesn't mean it should.

Figure 43. 432nd Wing technicians rebuild a Predator following its return from Afghanistan in 2008. The markings on the airframe signify missile launches. Air Force photo.

SWARMING

Industry has been refining the hardware, software and concepts for getting military robots to swarm like insects, but the most promising large-scale application for swarming—the Navy's Unmanned Combat Air System Demonstrator program—has been gutted by budget cuts. As a result, it might be years before a military customer combines available systems to make swarming a reality.

After several years of explosive proliferation, there are now thousands of Unmanned Aerial Vehicles in service with U.S. and allied forces. But the vast majority of them operate singly and with a human operator in the loop. Highly autonomous swarms offer many advantages. They can carry a wide number and variety of sensors and weapons—and a swarm is more redundant and less vulnerable than a single robot.

The Navy's $600-million Ucas-D program, which aims to test a fast, fighter-sized UAV, was the best opportunity to begin large-scale experimentation with an aerial drone swarm. But a week after the Navy picked Northrop Grumman's X-47 drone for the Ucas contract, the sea service announced that the six-year program would demonstrate only carrier operations. No weapons. No sensors. No swarms.

"The Navy has more or less told us in no uncertain terms that they are going to fund us to do the Ucas-D demonstration … and any future activities are undefined," said Bruce Parmenter from Northrop Grumman.

Economics and the program's troubled history were the major factors in the Navy's decision, Parmenter said. An earlier incarnation of the Ucas program included the Air Force as a partner and so had potential for more robust funding. But the Air Force dropped out in 2006 in order to devote more resources to new manned bombers. There was just enough money for a bare-bones Ucas demonstration.

Even so, industry is banking on the Navy eventually expressing a need for swarming. Several firms have been working on methods of getting large groups of drones to navigate together without colliding, to collaborate on tasks and to share data so that information isn't lost if one drone in the swarm is destroyed.

High-tech sensors aboard each robot, combined with sophisticated software algorithms to tell the drone to "sense and evade" another object, is one way to keep a swarm from destroying itself, but industry experts said that sense and evade is too expensive. Instead, several firms are working on networks that maintain distance between their nodes. These also offer the advantage of also being able to move imagery and targeting data from node to node.

Figure 44. An X-47 killer drone emerges from its production jigs. Northrop Grumman will use the jet-powered robot to test drone operations aboard aircraft carriers. X-47 might one day fly in large swarms. Northrop Grumman photo.

BUILD YOUR OWN WAR BOT

Swarming robot planes are quite sophisticated. But most military robots are actually very simple—so simple, in fact, that private citizens have come up with their own home-made versions. A combination of commercial radio-controlled hobby airplanes and toy trucks plus cheap cell-phone and digital camera technology and off-the-shelf software produces robots that, if not quite as capable as military robots, perform the same basic tasks.

The most popular military Unmanned Aerial Vehicle is the Raven. Raven's primary functions are to fly a short distance (a few tens of miles), look down and transmit what it sees. You can make your own Raven surrogate for cheap. The biggest difference your UAV and the real thing? Raven is fully autonomous: it follows GPS waypoints to its destination without human intervention. While R/C plane junkies are hard at work on do-it-yourself autopilots, most homemade drones are still remotely controlled. It's simpler, cheaper and safer.

If it flies, and can haul around 200 grams worth of equipment, it should work as a drone airframe. Amateur UAV operators have built their systems on R/C planes and helicopters, small balloons and even kites. Most of the platforms, being radio-controlled, are operated by way of handheld consoles and have ranges of just a few miles.

As far as sensors go, any point-and-shoot digital camera (set to snap photos on automatic) will work, but options for image processing in this case are limited. The latest camera phones boast superior image quality, video and the ability to stream imagery back to the operator in real time. These phones also boast integrated GPS, a requirement for some higher-end image processing.

With a basic digital camera or a stand-alone camera phone, image processing amounts to downloading the stored images or video onto your computer post-flight. If your sensor or platform also records GPS data, you can supply both the imagery and the GPS data to a company called Pict'Earth, and they'll overlay your images onto a Google Earth interface, for a fee. If you've got your own Pict'Earth hardware, you can receive the imagery in real time from the platform and build your own, improved, Google Earth maps.

You can simply duct-tape your sensor to your platform, with the facing camera aperture facing down, of course. With this method, it's necessary to start up the camera function before launching the platform, as there is no way to remotely activate the camera. More sophisticated installations are possible. Many R/C aircraft have belly bays for additional gear, often including GPS data loggers. Camera phones can be fitted in the bay alongside additional servos for remotely pressing buttons.

UAVs are, in many ways, easier than ground bots. In the air, there are few obstacles, and altitude offers an excellent vantage point. Ground robots, on the other hand, must contend with rough terrain and poor visibility. For those reasons, a practical do-it-yourself ground bot will remain fairly close to the operator—unlike a flying drone, which, with the installation of a forward-looking camera, might venture for miles.

Bomb disposal remains the most common mission for military ground bots. For that, a robot needs a camera for navigation and some means of delivering a payload (in wartime, this might be a clump of C4 for destroying a suspected bomb). The Army's most basic bomb-bot is the 25-pound, wheeled MarcBot, which is based on the chassis of a Traxxas E-Maxx toy, but any robust R/C truck will work for the DIY robot-maker.

Any small wireless camera setup will work for your sensor. One successful home-made robot-maker used a wireless camera atop radio-controlled tilt-and-pan servos. The servos allowed the camera to gaze independently of the robot's direction of travel.

Tackle Design's $1,000 OrdBot, which has seen (unofficial) action in Iraq, uses an R/C servo connected to a front ramp to deploy the payload, but there are any number of potential servo-based setups, including a ram for shoving the payload out of a tray. For your own bot, just strip down the R/C truck to its chassis, add cameras, servos and ramps as needed—and voila!

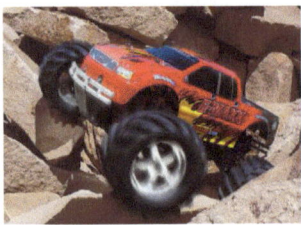

Figure 45. The Traxxas E-Maxx. Traxxas Photo.

THESE LEGS WERE MADE FOR THINKING

On April 5, 2003, Army Private First Class Garth Stewart stepped on a land mine in Baghdad. The explosion blew off his left leg beneath the knee. Like more than 500 military amputees from the Iraq war, today Garth wears a prosthesis to get around. As long as he's in blue jeans, you can't tell from looking at him. But he sure can tell.

Most of today's prosthetic lower-leg prostheses are essentially springs. They absorb shock, but don't push back the way a real leg does. That causes back pain and makes walking a chore.

But not for long, if one pioneering prosthesis designer has his way. Hugh Herr, an MIT professor who lost both of his legs below the knees in a childhood mountain-climbing accident, has developed many of the most advanced lower-body prostheses, including the popular Rheo knee with an embedded micro-processor. His latest, an intelligent robotic ankle, promises to make walking easier for Garth and others like him.

"Existing designs, when the foot's on the ground, are passive, and only provide as much energy as an amputee puts into them," Herr said in 2007. "The new device puts in added energy. It has its own energy source. The device can control its own stiffness and damping and resistance as the amputee walks on different terrains. An amputee can walk faster and with less energy."

The key, Herr said, was understanding how our biological legs—their joints, tendons and muscles—really work and translating that into plastic, metal and software. What's the hardest part? "Everything. It's a hard problem. We're trying to have a device that has three times the power of conventional prostheses, but has to be lightweight and small enough to scale down to a woman's size five."

The software was a challenge, too. There were robotic prostheses already in use, but they were too dumb to react quickly to changing terrain. There was latency: the limb didn't know that you were climbing stairs, for instance, until after you'd taken a couple steps up. Herr's new ankle requires a smarter computer brain in order to detect changes seemingly instantaneously.

After months of tinkering, in March Herr began testing "the device," as he and his assistants referred to the ankle, in his lab at MIT. Garth was one of his subjects, selected because he lived nearby and because he was heavier, more athletic and more comfortable in a prosthesis than most amputees. Garth did two daylong sessions at MIT, walking 40 feet at a time, providing feedback while a technician modified the ankle's software algorithms in real time. Herr himself also wore two of the experimental ankles to generate more test data. By July 2007 their confidence in the design was such that MIT and the ankle's co-financiers, Brown University and the Providence V.A. Medical Center in Rhode Island, hosted an official unveiling. By then "the device" had a name: iWalk.

All this robotics work is making amputees' lives easier, but the benefits don't end there. According to Herr, advancements in prosthetics apply directly to other fields, including more traditional robotics. Building a robot with man-shaped legs, for instance, requires many of the same technologies as attaching a leg-shaped robot to a man. Research into prostheses might one day result in robots with the same strength and dexterity as human beings—robots that can run, jump and climb.

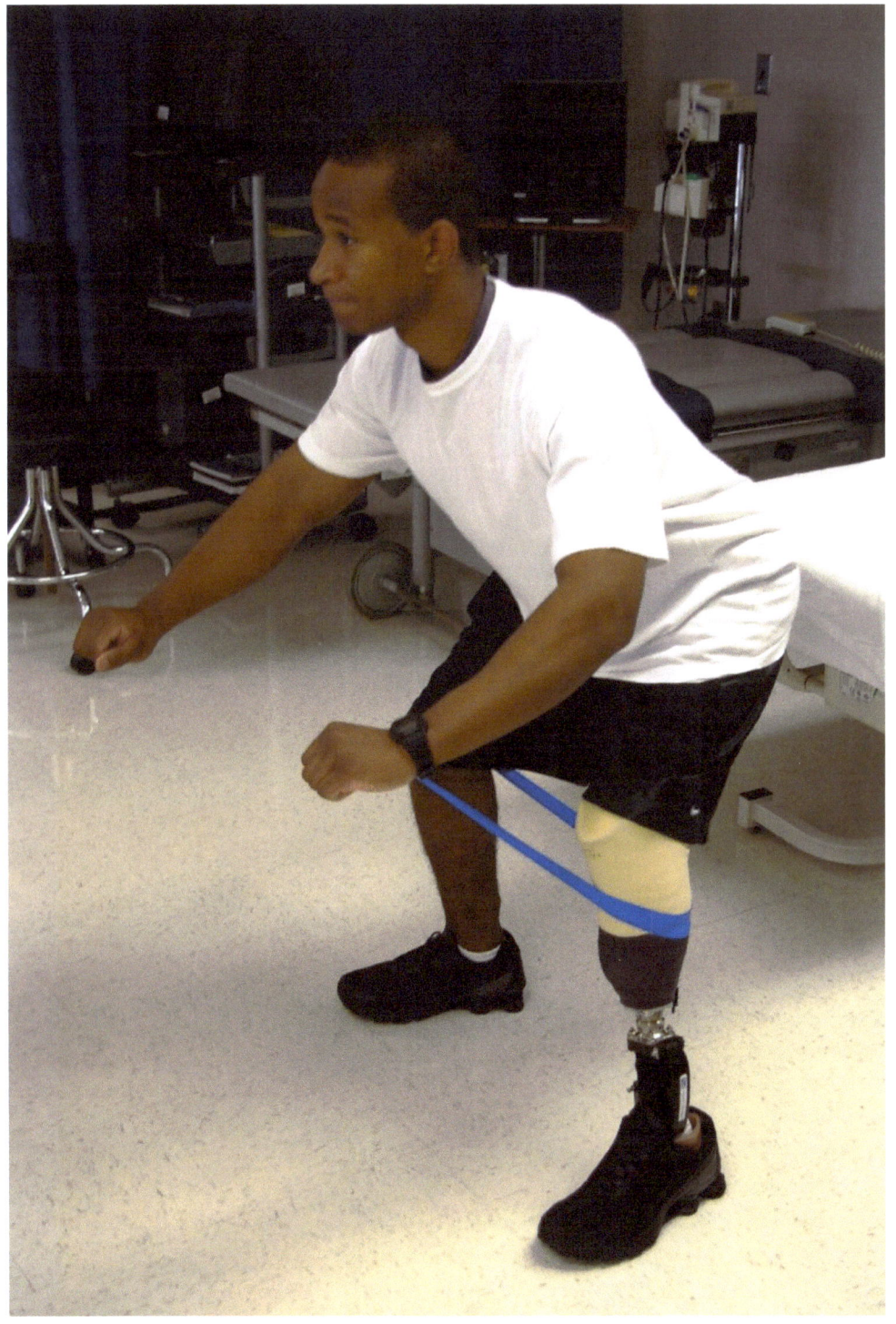

Figure 46. Hundreds of amputees from the wars in Iraq and Afghanistan have received robotic prostheses from the Veterans Administration. Photo by David Axe.

Figure 47. Robotic ankles detect changes in the terrain and adjust accordingly. Hugh Herr's new iWalk ankle also duplicates the "push-back" effect of real muscles. Photo by David Axe.

Figure 48. Garth Stewart, a former soldier, tested the first iWalks with designer Hugh Herr. Robotic prostheses might one day lead to truly humanoid robots. MIT photo.

Swimming Lessons

It's one thing to get a robot to fly through the air or crawl or walk around on the ground. The air is mostly devoid of obstacles. On the ground, the obstacles are pretty numerous but at least you're only moving on roughly one plane. Underwater robots, on the other hand, are literally submerged in an obstacle.

Being underwater makes robot operations very difficult, according to BAE Systems engineer Norm Scott, not the least because you can't send a command signal through water. So most current underwater robots go out on short missions then come back to a ship to talk to their operators. BAE's Talisman, which the company promoted in 2007, aims to change all that. Talisman is a bigger, tougher and more flexible robot based on racecar technology. It's intended primarily for detecting and destroying underwater mines.

"Existing Unmanned Underwater Vehicles have the capability to detect mines but have to go back to the ship and download their information so the ship can see what the UUV sees," Scott said. But Talisman just surfaces, sends out a radio signal and waits for a reply.

Talisman looks like a sleek window- and wheel-less compact car— and in fact is built from the same materials as modern racecars. It has four little jets on the bottom and a weapons bay for carrying torpedoes, smaller robots or charges for destroying mines. Scott said Talisman has more horsepower than most undersea bots and can make headway against a five-knot current.

In light of the challenges facing underwater robots, Talisman faces an uphill climb to widespread acceptance. Today the robot is still just a demonstrator. But BAE has faith in the sleek robot, and in 2007 took it on a road trip to show it off at U.S. Navy installations on the East and West Coasts.

Underwater is the next frontier for military robots.

Figure 49. A sailor deploys a mine-hunting robot from the side of a destroyer in 2007. Underwater is the toughest environment for robots, since they cannot reliably send and receive transmissions. Navy photo.

Postscript

Law Zero: A robot may not injure humanity, or, through inaction, allow humanity to come to harm.

Law One: A robot may not injure a human being, or, through inaction, allow a human being to come to harm, unless this would violate a higher order law.

Law Two: A robot must obey orders given it by human beings, except where such orders would conflict with a higher order law.

Law Three: A robot must protect its own existence as long as such protection does not conflict with a higher order law.

OTHER TITLES ABOUT MODERN WAR FROM NIMBLE BOOKS

You Can Run But You Can't Hide: European UAVs by Tim Mahon (2009)

CV-F THE FUTURE UK AIRCRAFT CARRIER by Tim Mahon (2009)

The John Boyd Roundtable edited by Mark Safranski (Zenpundit)

Revolutionary Strategies in Early Christianity by Dan H. Abbott (tdaxp)

CVN-68 NIMITZ, U.S. Navy Aircraft Carrier

CVN-69 DWIGHT D. EISENHOWER, U.S. Navy Aircraft Carrier

CVN-70 CARL VINSON, U.S. Navy Aircraft Carrier

CVN-71 THEODORE ROOSEVELT, U.S. Navy Aircraft Carrier]

CVN-72 ABRAHAM LINCOLN, U.S. Navy Aircraft Carrier

CVN-73 GEORGE WASHINGTON, U.S. Navy Aircraft Carrier

CVN-74 JOHN C. STENNIS, U.S. NAVY AIRCRAFT CARRIER

CVN-75 HARRY S TRUMAN **(2008)**

CVN-76 RONALD REAGAN, U.S. Navy Aircraft Carrier

CVN-77 GEORGE H. W. BUSH, U.S. Navy Aircraft Carrier

CVN-78 GERALD R. FORD, U.S. Navy Aircraft Carrier

DDG-81 WINSTON S. CHURCHILL, U.S. Navy Destroyer]

SSN-23 JIMMY CARTER, U.S. Navy Submarine (Seawolf class)

X-47 Unmanned Combat Air Vehicle (UCAV)

To purchase, visit http://www.nimblebooks.com.

Publish a "Nimble" Book

Did you enjoy this book? If so, how about writing something similar for us! We have series on Modern Weapons, Modern Warships, World War II, Modern Aircraft, and Napoleonics. Other ideas are of course quite welcome.

The sad truth is that many (if not) most full-length books today are padded. Sometimes it can be more artistically effective to work within constraints. The manuscript should be 7,000 to 32,000 words long minus approximately 400 words for each full-page image. Excluding the front and back matter, that works out to 23 to 120 pages for the body of the manuscript (including images, which are typically presented in full-page format).

Images must be 72dpi resolution or greater; color, if at all possible; and provided with either explicit permission to use from the owner or with a strong "no copyright" rationale.

To be clear, the availability of color images is not a requirement; it's a nice-to-have. This format does work with straight text or with B&W illustrations, but color helps differentiate the book and add value for the customer.

The "content package" that seems to motivate purchase best is a substantial number of beautiful images plus pointed, insightful commentary on a topic that is attractive to enthusiasts.

The important thing to understand about the financial side is that this is micropublishing. I will be quite happy with a book that sells 10 copies a month. This is not a way to get rich; I would suggest that you think about it more like adding a revenue stream to your blog. By that measure this approach stacks up pretty well. Each book purchased (anywhere in the world) is like a $3 ad click. There are no fees to the author, ever, and we bear all costs of publication.

Incidentally, I have no objection if you wish to compose your book as a series of blog entries, then turn a manuscript over to me for "nimble" publication. Our standard contract gives Nimble exclusive rights to publication in book form, but all other rights are reserved to the author.

This format should work for you if:

- You have pointed, interesting text in hand that you want to "repurpose" or try out in print, or
- You have beautiful color photos or art in hand, or
- You pick a topic that is easily illustrated with high-quality public domain images (e.g. space, navy ships, history before 1922)

and ...

- You have a blog or some sort of platform that gives you the ability to drive traffic or...
- You pick a topic that is keyword-friendly or ...
- You pick a topic that is beloved by enthusiasts.

The Nimble Books Marketing Playbook provides advice on how to maximize your online sales.

A proposal form is available on our web site (www.nimblebooks.com) under "Publish With Us."

Letter from the Publisher

Dear Reader,

The introduction of robots into war may ultimately be as important as the introduction of the stirrup, the longbow, gunpowder, the rifle, the machine gun, the tank, the aircraft, and the atomic bomb. Where the atomic bomb moved mass casualties into the realm of the "unthinkable," the war bot may move even individual casualties into the realm of the "avoidable"—for the richest nations.

I am proud that David Axe has brought this brilliant collection of beautiful images and pointed text to Nimble Books. It captures a critical moment in the history of military technology. Fifty years from now, I can hope, looking back at this book will be like looking back at a contemporary album of the Sopwith Camels and the Fokker Triplanes. In the meantime, it is essential reading for anyone who wants to know the changing face of war.

—*Fred Zimmerman, Nimble Books LLC,*
Ann Arbor, Michigan, USA, 2008

Figure 50. The Sopwith Camel (left) and the Fokker Dr I (right). Wikimedia Commons photo.

www.ingramcontent.com/pod-product-compliance
Lightning Source LLC
Chambersburg PA
CBHW041542220426
43664CB00002B/28